Middle School Science

Physical and Chemical Changes Science Workbook

This book is dedicated to the students I worked with at Rogers. I am forever thankful for each of you, our time together, and what you all taught me.

The clipart, images, digital papers, and fonts featured in this book are from OldMarket, CarrieStephensArt, Bricks and Border, Kimberly Geswein Fonts, Ron Leishman Digital Toonage, Janie Manning, and myself.
https://www.teacherspayteachers.com/store/oldmarket
https://www.teacherspayteachers.com/store/carriestephensart
https://www.teacherspayteachers.com/store/bricks-and-border
https://www.teacherspayteachers.com/store/kimberly-geswein-fonts
https://www.teacherspayteachers.com/store/ron-leishman-digital-toonage
https://www.teacherspayteachers.com/store/elly-thorsen

Table of Contents

Instructions for Use

Overview of Contents:

This workbook contains a variety of materials to learn, practice, and assess the differences between physical changes and chemical changes. A reading passage provides the information needed to master physical and chemical changes. Worksheets, activities, experiments, a project, and an assessment are included. Detailed answer keys are in the back of the workbook.

Suggested Implementation:

Begin by having your student read the informational text about physical and chemical changes. Then complete the worksheets and activities that follow. Alternating between worksheets and activities will help reinforce the material without it becoming tedious or overwhelming.

A wide variety of worksheets are included, and they are arranged in order from easiest to most challenging. Your student can complete all of the worksheets or you can pick and choose which worksheets to prioritize.

Multiple activities are included and all of them have detailed instructions for use. The first is a card sorting activity. The second activity is a set of stations that can be used as task cards or be posted around the room for your student to visit. There is an ideas list of hands-on activities you can do with your student to experience physical and chemical changes. Two experiments requiring minimal and easy-to-find materials round out the activities.

End your physical and chemical changes unit with the poster project, written assessment, or both.

Suggested Pacing Guide:

- **Day One:** Complete page 5. Read pages 6-8. Complete page 10.
- **Day Two:** Review using page 9. Complete pages 11-12.
- **Day Three:** Do the first half of the Egg Experiment on page 38. Complete page 13. Choose a hands-on activity from page 40.
- **Day Four:** Complete the card sorting activity on pages 17-25.
- **Day Five:** Complete the rest of the Egg Experiment on page 38. Complete page 14. Choose a hands-on activity from page 40.
- **Day Six:** Complete the stations activity on pages 26-36.
- **Day Seven:** Do the Balloon Experiment on page 39. Complete page 15.
- **Day Eight:** Complete page 16. Choose a hands-on activity from page 40.
- **Day Nine:** Begin working on the poster project on pages 41-42.
- **Day Ten:** Finish the poster from Day Nine. Complete the unit with the assessment on page 43.

Physical and Chemical Changes: An Introduction

What do you know about physical and chemical changes?

- In the space below, use sentences and pictures to record everything you already know about physical and chemical changes.

Physical and Chemical Changes: An Introduction

Matter Changes

Imagine a bag of jumbo marshmallows. You fish a marshmallow out of the bag and smash it into a pancake before chewing it up and swallowing it. You take another marshmallow, shove it on a stick and hold it over a fire, watching it puff and blister before turning black and crispy. You melt the rest of the marshmallows together with butter and mix it all together with cereal to make a delicious snack. Why all the talk about marshmallows? Marshmallows are matter, and matter can change.

Key Terms

Mass: The amount of stuff in an object

Volume: The amount of space an object takes up

Atom: The smallest possible piece of matter

What is Matter?

Matter is anything that has mass and volume. Matter is made of tiny particles called atoms. Matter can go through two categories of changes: physical changes and chemical changes.

Physical Changes

Physical changes are changes that do not change the identity of a substance. The size, shape, or texture of the object might have altered but not its atomic makeup. In other words, the item is still itself—it might just look or feel different. Remember that marshmallow you smashed into a pancake? Totally a physical change.

Changes in states of matter, such as melting and freezing, are always physical changes. For example, if an ice cube melts it changes from a solid state to a liquid state but it is still water. Another example of a physical change is cutting paper. The paper is now in several pieces, but it did not turn into another substance. Oftentimes, physical changes can be easily reversed.

Physical and Chemical Changes: An Introduction

Chemical Changes

A chemical change causes the identity of a substance to change. The item is no longer the same thing it was before the change. Something new is created that was not there before. Chemical changes involve rearranging atoms. New substances are created during chemical changes because the bonds holding atoms together are broken and new bonds are created. Remember roasting that marshmallow? The black crust that formed on the outside certainly was not there before you held the marshmallow over the fire. The marshmallow went through a chemical change.

Chemical changes can happen when items are burning, being digested, or reacting with other things. There are many signs that indicate a chemical change has occurred.

- The production of light, heat, smoke, or gas is a sign that a chemical change likely occurred.
- A change in color often points to a chemical change.
- A new smell also means a chemical change might have happened.
- The formation of a precipitate is another sign of a chemical change, though one you are less likely to see frequently in your everyday life. (A precipitate is a solid that forms when two liquids are mixed together.)

An example of a chemical change is the banana you bought weeks ago that is now rotting on the counter. The banana gradually turned darker and smellier, going from a nice yellow banana with a mildly sweet smell to a black banana with a putrid stench. What are the signs of a chemical change in your banana? That's right, color change and a new smell. Chemical changes are very difficult or even impossible to reverse. It is not possible to un-rot that banana.

Physical and Chemical Changes: An Introduction

Telling the Changes Apart

To distinguish between a physical change and a chemical change, first look for the signs of a chemical change. Is there a change in color or a new smell? Do you notice smoke, a temperature change, or light? Do you see any bubbles forming or any new solids at the bottom of a liquid? Sometimes the answer might not be obvious, so you have to dig a little deeper and ask more questions. Is the change easily reversible? Do I have something new that wasn't there before?

Ask Yourself:

Do I have something new that was not there before?

- If no → It is probably a physical change.
- If yes → It is probably a chemical change.

Physical Change Example

After an intense storm there are often fallen trees on roads and branches scattered all over. In order to clean up the trees they must be cut into more manageable pieces of wood. The tree gets smaller with each swing of the axe, but it is still wood. If you put the tree branches into a woodchipper, then you have tiny pieces of wood. No new substance was made that was not already there. The tree looks and feels entirely different but it is still wood. This is a physical change.

Chemical Change Example

Instead of putting the fallen trees through a woodchipper, what if you were to have a bonfire instead? As you stand near the burning branches you can see smoke and light from the fire, feel heat, and notice a color change. Once the fire dies down, you have a new substance—ash. This is a chemical change.

Your Guide to Physical and Chemical Changes

Physical	Chemical
The particles in the item stay the same. Nothing new is made. You still have the same materials you started with. It is often easy to change the item back to how it used to be. Examples: ○ Mixing cake batter ○ Freezing water ○ Cutting paper	The atomic makeup of the item changes. Something new is created that was not there before. It is difficult to change the item back to how it used to be. Signs include changes in light, heat, gas, smell, and color. Examples: ○ Baking a cake ○ Milk turning sour ○ An old pan rusting

Physical and Chemical Changes Overview

1. Write the definition of a physical change in your own words. ______________________________
____
____

2. List two examples of physical changes. ______________________________
____

3. Write the definition of a chemical change in your own words. ______________________________
____
____

4. List two examples of chemical changes. ______________________________
____

5. Are changes in states of matter, such as evaporation and freezing, examples of physical or chemical changes? Explain your answer. ______________________________
____
____

6. What are some signs a chemical change occurred? ______________________________
____
____
____

7. What are some questions you can ask yourself to help you determine if a change is physical or chemical? ______________________________
____
____
____
____

In each box below, draw one way you can change a sheet of paper. Then label your drawings as physical or chemical changes.

Physical and Chemical Changes Makeover

Directions:

It's time for a makeover! In the bottom photo, draw how you look now. Then, in the top photo, draw how you imagine you might look after experiencing the physical and chemical changes your stylist or barber put you through. Make sure to color your drawings. List all of the physical and chemical changes you experienced.

After my makeover

Before my makeover

Physical Changes:

Chemical Changes:

Physical and Chemical Changes Examples

Directions: Circle the physical changes and underline the chemical changes.

1. A plate breaks in half on the floor.

2. You notice the leaves change color from green to orange in autumn.

3. Your little sister burns toast in the toaster.

4. Your hot shower causes condensation to develop on the mirror.

5. You mix flavored drink powder with water.

Directions: Determine whether the changes are physical or chemical and then explain how you know what kind of change occurred. Answer in sentences.

6. Eating a chocolate cake donut with rainbow sprinkles: ______________________________

____

____

____

7. Strawberries turning from green to red as they grow: ______________________________

____

____

____

8. An aluminum can crushing under your boot: ______________________________

____

____

____

9. Milk turning sour after its expiration date: ______________________________

____

____

____

10. A popsicle melting on a hot sidewalk: ______________________________

____

____

____

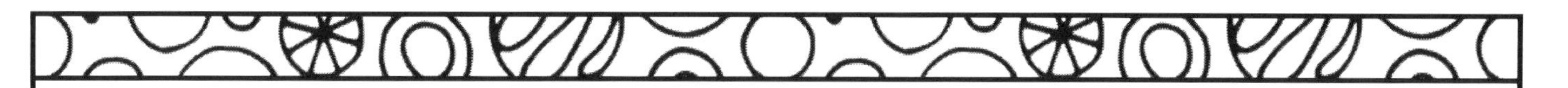

Physical and Chemical Changes Identification

Directions: Read each paragraph. Identify the change as either physical or chemical. Explain your answer in paragraph form.

1. Last summer you watched a fireworks show with your family. The fireworks were so colorful and bright. Each explosion lit up the black sky. When the wind drifted over to where you were sitting you smelled smoke from the fireworks.

____
____
____
____
____
____
____
____
____

2. Your mom asked you to babysit your little sister for a few hours. You decided to make paper snowflakes together. You each folded a piece of paper and used scissors to make small cuts along the edges. When you unfolded your paper, you had a beautiful and unique snowflake. When your sister unfolded her paper, she had an interesting-looking hacked apart piece of paper.

____
____
____
____
____
____
____
____
____

3. Yesterday at lunch you ate a hamburger. A few hours later your stomach started to gurgle and you began to feel uncomfortable with all the gas moving around in your belly. This morning yesterday's lunch reemerged while you were in the bathroom. Your hamburger looked and smelled very different. Also, it was a new color.

____
____
____
____
____
____
____
____
____

Physical and Chemical Changes in Your Daily Life

Directions: For each location, give one example of a physical change and one example of a chemical change. Then explain in a sentence how you know the change is physical or chemical.

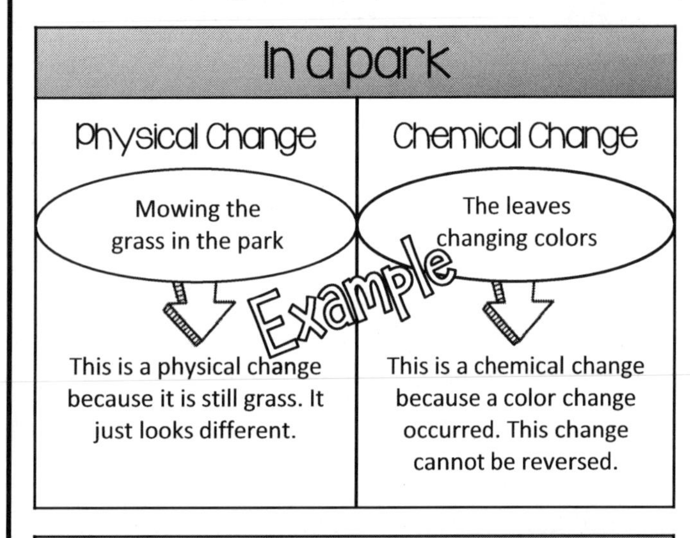

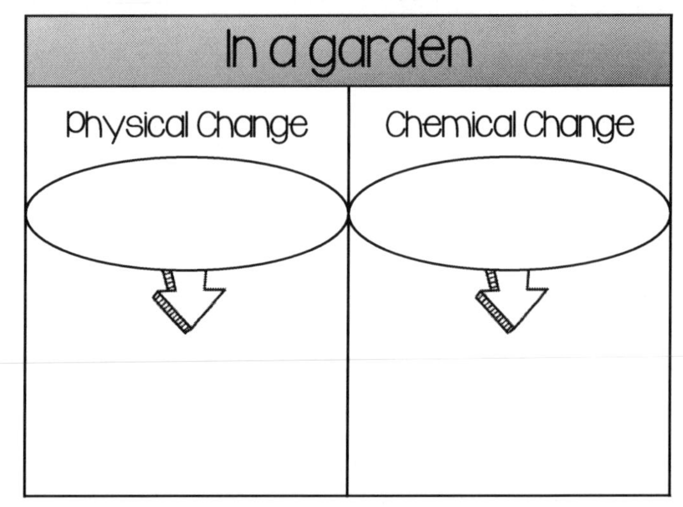

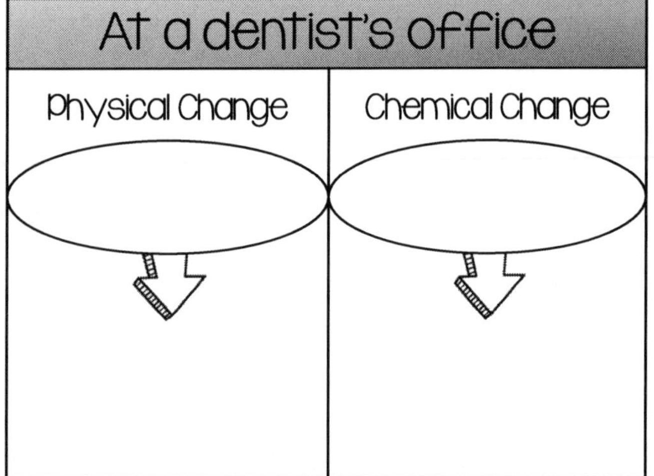

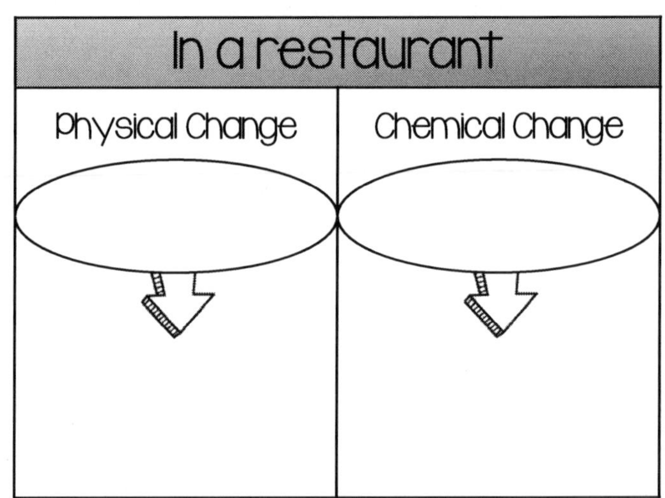

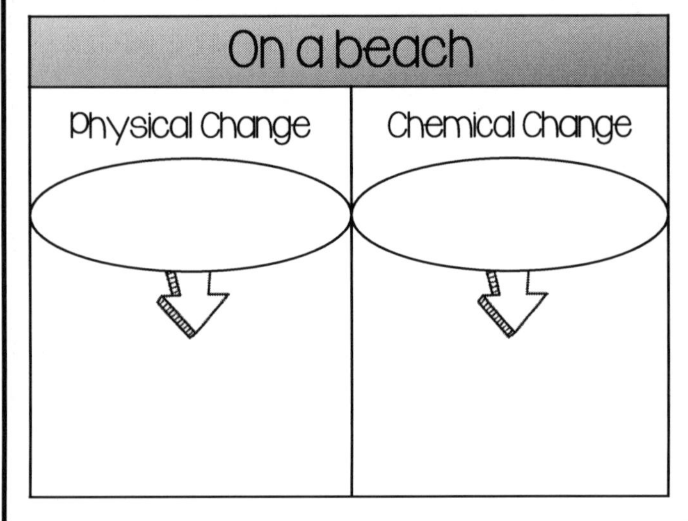

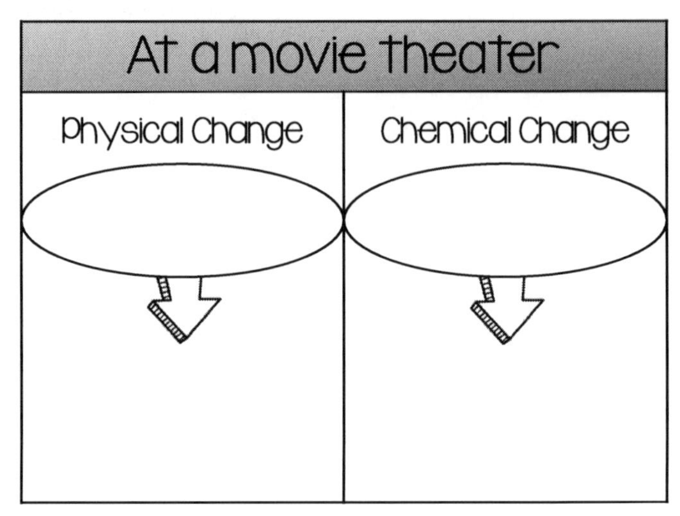

Physical and Chemical Changes Write or Draw

Directions: Complete each section below with words and/or drawings.

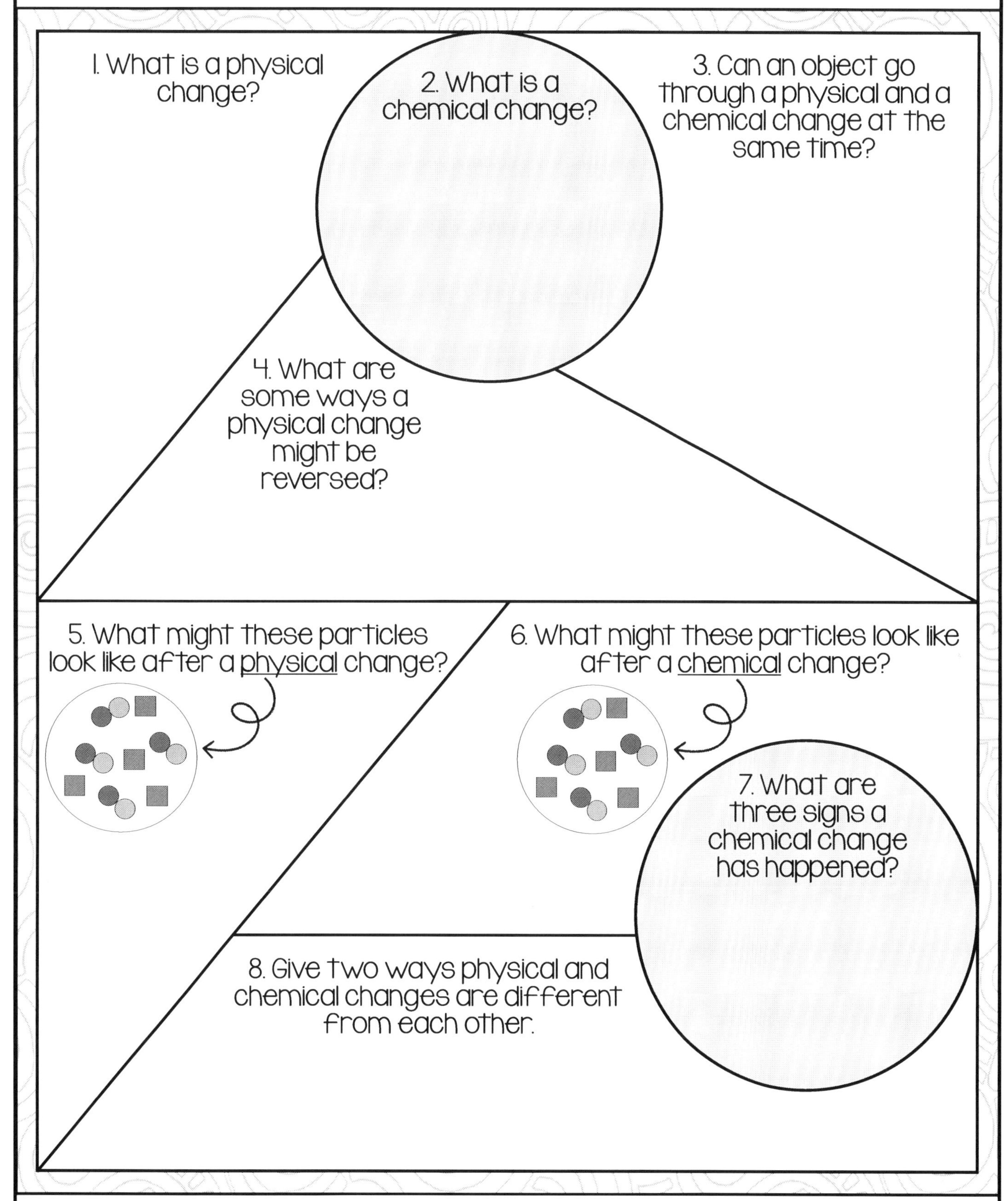

Physical and Chemical Changes Odd One Out

Directions: Analyze each part of the circle. One of the parts will be different than the others. Shade the part of the circle that does not match. Using a complete sentence, explain how that part is different from the others. *Answers must be science related and should show your understanding of physical and chemical changes.*

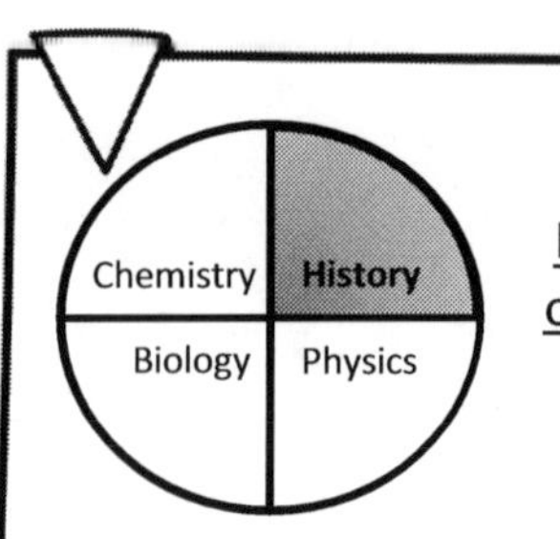

Example

History is the odd one out because it is not a type of science like the others.

1

The item is still itself.	Something new is created.
The item can change state.	The change can often be reversed.

____

2

Burning a log in a campfire	Baking cookies
Breaking a glass	Digesting food

____

3

A change in shape	The production of light
A new smell	The creation of heat

____

4

The identity of the item is different than before.	A new substance is formed.
The bonds holding atoms together are broken.	The identity of the item is the same as before.

____

5

Boiling water	Freezing water
Dissolving salt in water	Making water by joining oxygen and hydrogen atoms

____

6

A paper turns from white to red with paint	Leaves change color in the fall
A flamingo turns pink from its diet	A person's skin tans in the sun

____

Card Sorting Activity

What will my student do in this activity?

Your student will sort cards into groups of physical changes and chemical changes. After your student sorts the cards, they can complete the chart and answer the questions on their Card Sorting Activity Sheet on page 25.

When should I use this activity?

Once your student understands the definitions and is familiar with the signs of chemical changes, they should be ready for this activity. This activity can be used any time after reading the Physical and Chemical Changes Introduction on pages 6-8.

Directions:

1. Cut out the cards on pages 19-24.
2. Have your student follow the directions on the Card Sorting Activity Sheet (page 25) to complete the activity.
3. During the activity you can monitor your student and check their answers to make sure they are on the right track.
4. After your student is finished grouping the cards, you can go over the correct classification together. Your student should explain how they decided whether a card was a physical or a chemical change.
5. Your student should then complete the rest of the Card Sorting Activity Sheet by themselves.

Extension Idea:

Your student can use the blank cards on page 23 to make their own example cards with pictures.

Directions: Cut out each of the cards.

A change that does not change the identity of a substance; it is still the same but might look or feel different

A change that changes the identity of a substance; something new is created that was not there before

Directions: Cut out each of the cards.

Directions: Cut out each of the cards.

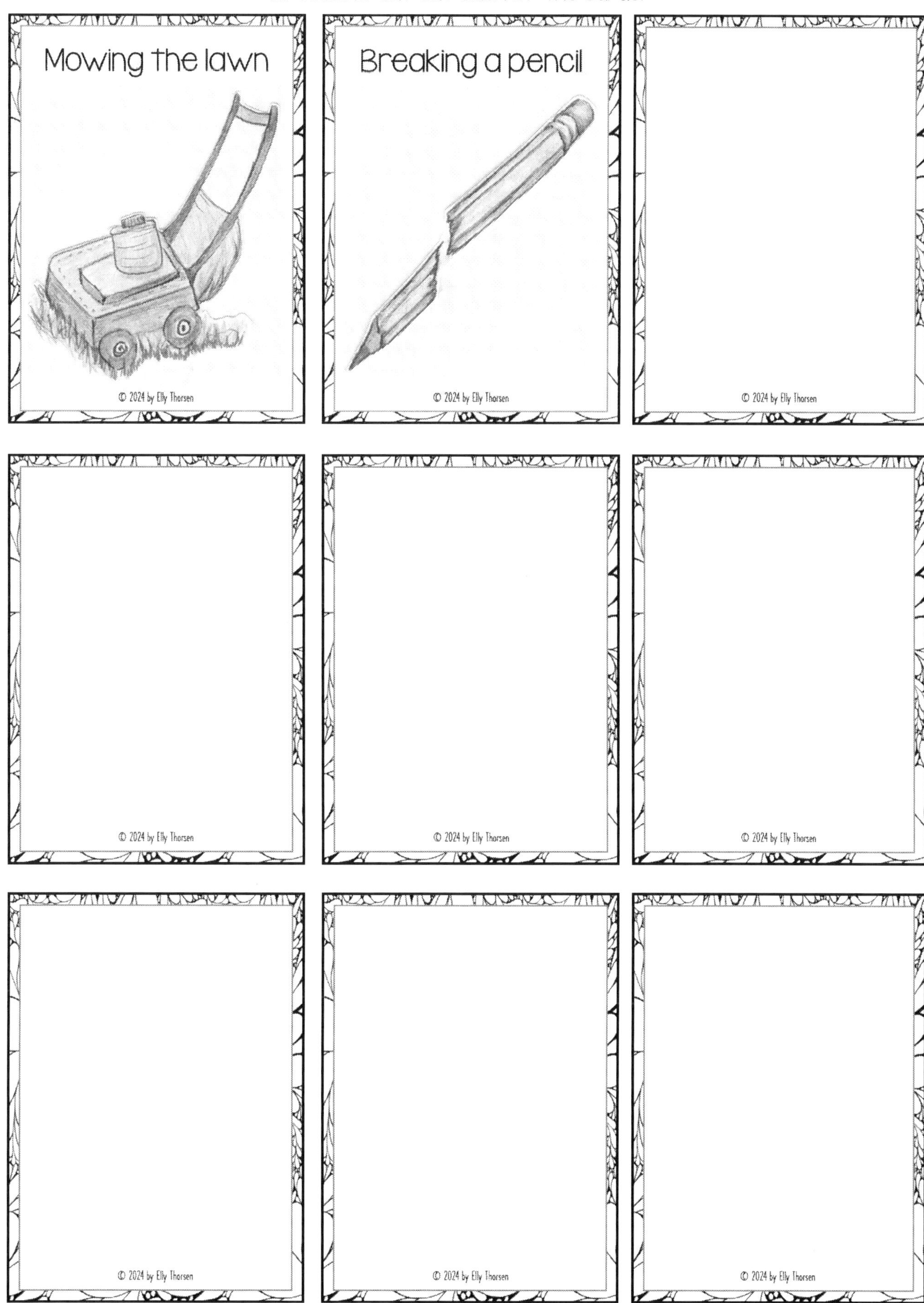

Card Sorting Activity Sheet

Directions:

1. Lay the Physical Change card and Chemical Change card in two separate piles.
2. Match the definition cards to the correct type of change.
3. Match the remaining cards into examples of physical and chemical changes.
4. Check your answers.
5. Record the correct answers in the chart below.
6. Complete the rest of this activity sheet.

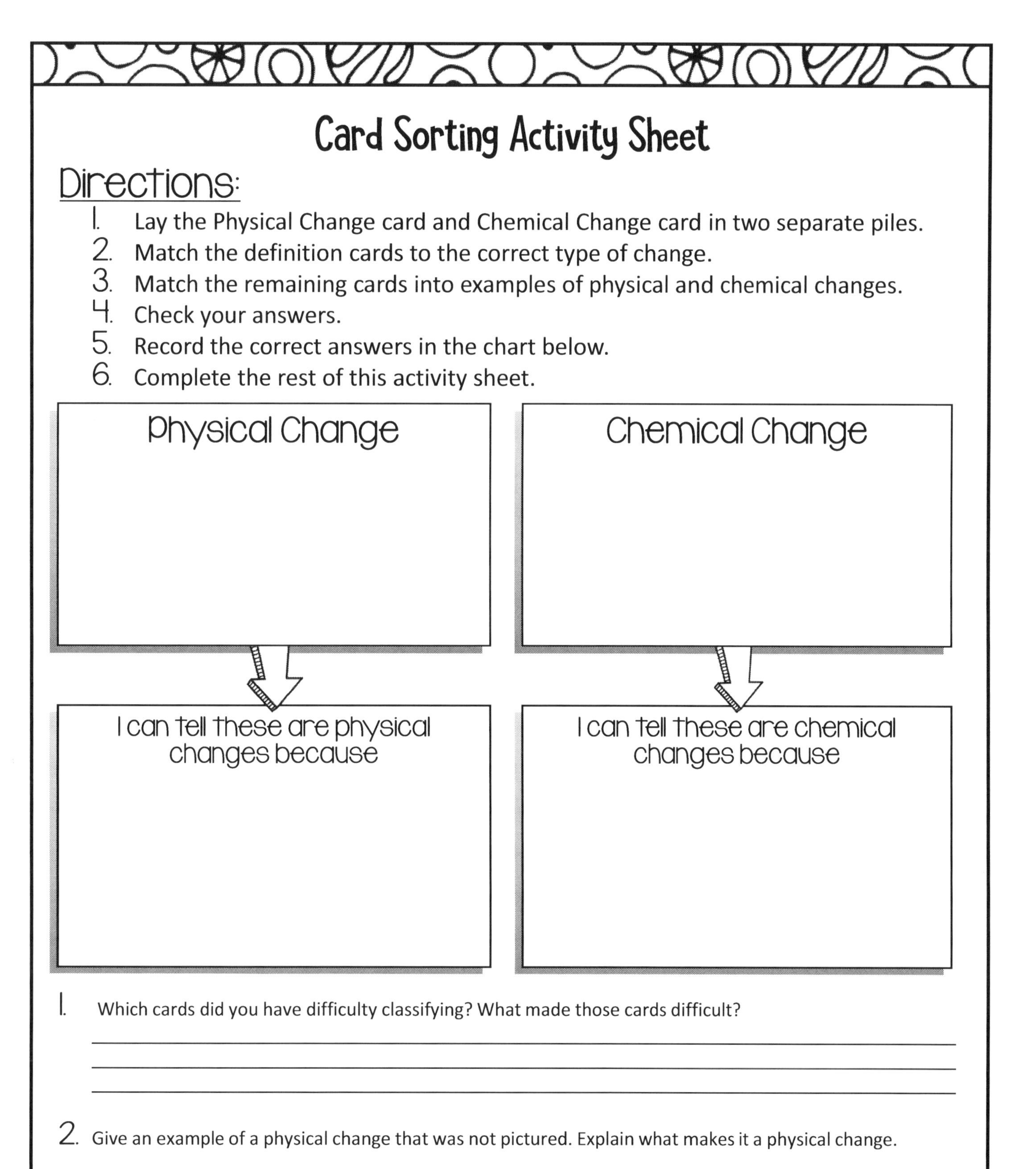

1. Which cards did you have difficulty classifying? What made those cards difficult?

____

____

____

2. Give an example of a physical change that was not pictured. Explain what makes it a physical change.

____

____

____

3. Give an example of a chemical change that was not pictured. Explain what makes it a chemical change.

____

____

____

Stations Activity

How does my student complete this activity?

There are three options for how to use this activity.

- Option 1: Cut out the station cards on pages 27-36 and post the cards around a given area. You can hide the stations and do a scavenger hunt for the cards. Or you can tape them around the room, up and down a hallway, or outside. Your student should find or visit each station and write their answers directly on each card.
- Option 2: Cut out the station cards on pages 27-36 and stack them in a pile. Use the cards together with your student like you would flashcards. Or have your student complete the cards on their own by writing their answers directly on the cards.
- Option 3: Do not cut out the cards. Instead, complete the cards exactly how they are in this workbook.

When should I use this activity?

Once your student understands the definitions and is familiar with the signs of chemical changes, they should be ready for this activity. This activity can be used any time after reading the Physical and Chemical Changes Introduction on pages 6-8.

Extension Idea:

Your student can use the blank cards on page 35 to make their own example cards.

What is the definition of a physical change?

1.

What is the definition of a chemical change?

2.

Directions: Cut out each of the cards.

True or False: Many chemical changes can be easily reversed.

(Then defend your answer in a sentence.)

3.

True or False: A change in state of matter is a physical change.

(Then defend your answer in a sentence.)

4.

List six signs that a chemical change has occurred.

5.

© 2024 by Elly Thorsen

Which of the changes listed below are physical changes?

Stirring sugar into tea

Grilling hamburgers

Mixing brownie batter

Spreading butter on bread

6.

© 2024 by Elly Thorsen

Directions: Cut out each of the cards.

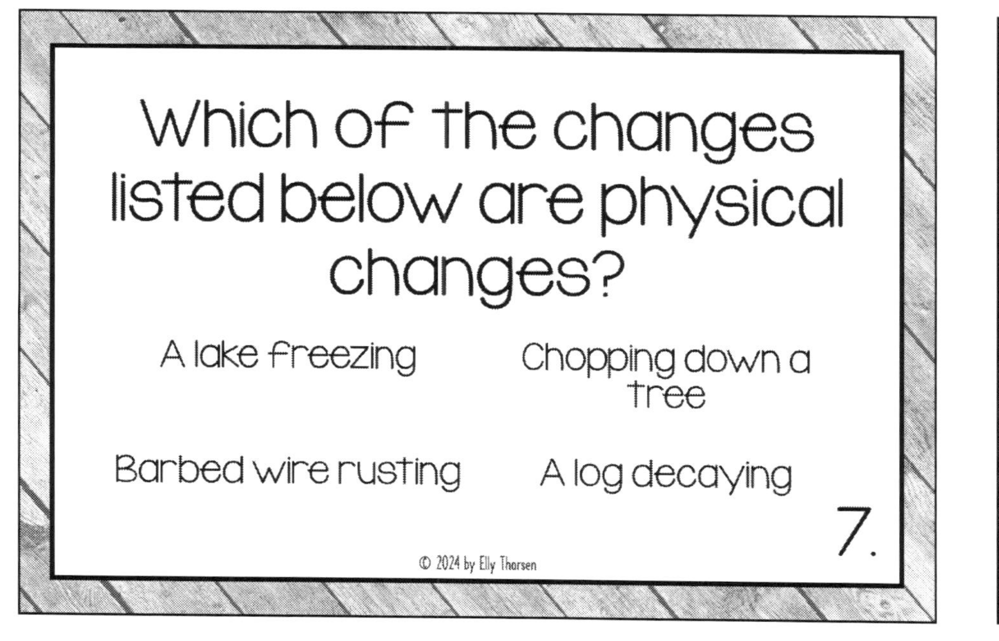

Draw a plant going through a chemical change.

8.

© 2024 by Elly Thorsen

Draw a plant going through a physical change.

9.

List three ways you can chemically change a carrot.

10.

Directions: Cut out each of the cards.

List three ways you can physically change a carrot.

11.

Which of the changes listed below is NOT a physical change?

- Melting chocolate
- Steaming up the bathroom
- Absorbing water with a sponge
- Freezing ice cubes
- Rising bread dough in a bowl
- Drying wet clothes

12.

Directions: Cut out each of the cards.

Make a list of eight physical changes.

14.

Read the paragraph and list the chemical changes.

Last Saturday you and a friend were playing video games. You decided to make popcorn and put a bag in the microwave. Soon you heard the corn popping. You were in an intense battle in the game, so you kind of forgot about the popcorn until you smelled smoke. The bag was on fire. After putting out the fire, you fed the burnt popcorn to your dog, Lady, who spent the rest of the day emitting rancid toots.

15.

Read the paragraph and list the physical changes.

Your little brother is making a card for your aunt's birthday. He cuts paper into a heart shape and colors the heart pink, yellow, and blue. Then he glues on fuzzy pompoms. After that he folds the heart in half and stuffs it into an envelope. He gives the card to your aunt and runs to the kitchen where he promptly eats the entire birthday cake "by accident."

16.

Directions: Cut out each of the cards.

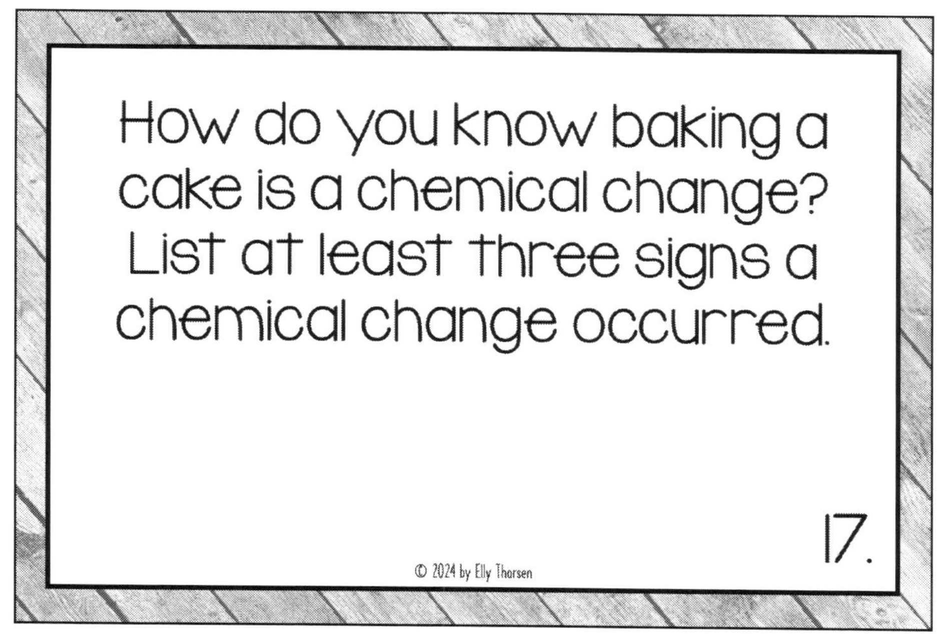

How do you know baking a cake is a chemical change? List at least three signs a chemical change occurred.

17.

How do you know a rotting pumpkin is a chemical change? List at least three signs a chemical change occurred.

18.

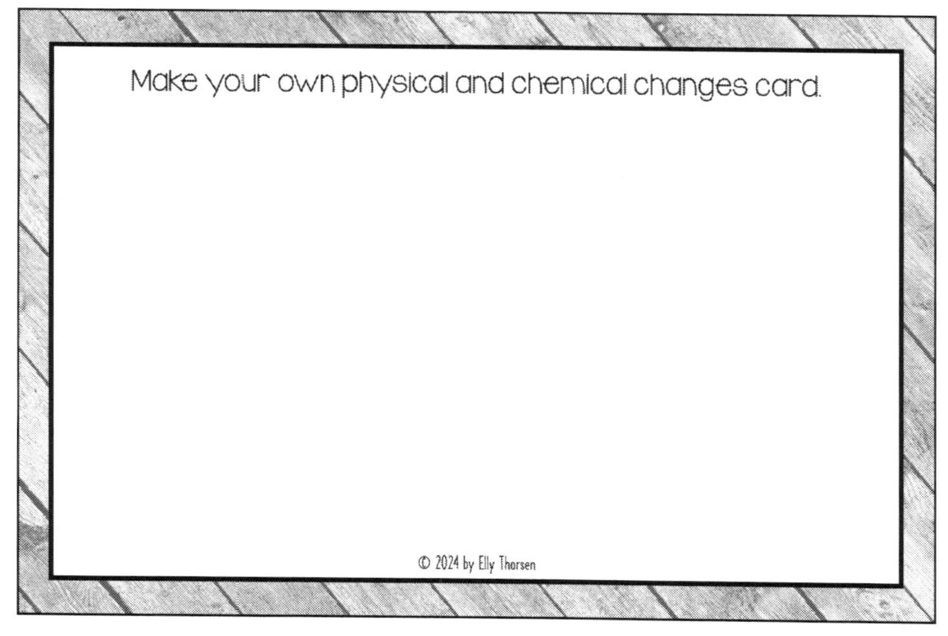

Make your own physical and chemical changes card.

Make your own physical and chemical changes card.

Overviews of the Experiments

Egg Experiment

- For this experiment you need an egg, a container like a cup or jar, and white vinegar. Food coloring is optional. Place an egg in the container. Pour vinegar over the egg until the egg is *completely* submerged. If you are using food coloring, add it to the container now. Observe the egg; you should start to see bubbles forming on the outside of the shell. After a full 48 hours, gently rinse the egg in water and carefully rub away the coating with your fingers. Notice how the egg looks and feels different. Your student can play with the egg, shine a flashlight on the egg to see the yolk inside, and bounce it. *Beware: the egg can and likely will pop, so it is best to do the egg handling outside or in an area that is easy to clean.*
  - This is a chemical change. The vinegar and the calcium carbonate in the eggshell react. You can see a sign of the chemical change in the form of carbon dioxide gas bubbles. The eggshell gets dissolved and leaves behind the thin membrane surrounding the egg.

Balloon Experiment

- This is one that is always seen at a science fair in the form of erupting volcanoes. You can take that concept a step farther by adding a balloon and an empty drink bottle. Start by pouring approximately a tablespoon (15 grams) of baking soda into a balloon. A funnel works nicely for this. Do not tie the balloon. Add approximately 4 ounces (120 milliliters) of white vinegar to a clean plastic bottle. Keep the lid off. Carefully stretch the lip of the balloon over the neck of the bottle, making sure the baking soda does not spill into the bottle. Once you have the balloon sealed on the bottle you can tip the contents of the balloon into the bottle. Keep your fingers pinching the area where the balloon and bottle meet so the balloon does not pop off. You will see and hear the vinegar and baking soda bubbling together. Gently swirling the bottle helps them combine to create a gas that will cause the balloon to expand.
  - This is a chemical change. When the vinegar (an acid) and the baking soda (a base) combine they create carbon dioxide gas which makes the balloon expand.
  - Keep experimenting: Have your student try different amounts of baking soda and vinegar to see what happens.

Egg Experiment

Needed Materials:

- One raw egg straight from the carton (check for cracks!)
- One clear container or jar
- White vinegar
- Food coloring (optional)

Predict:

In this experiment you will place an egg in vinegar for 48 hours. Make a hypothesis to predict what will happen to the egg. You can use the provided sentence frame to help organize your thoughts.

My Hypothesis

If I soak an egg in vinegar for 48 hours, then ______________

____

(explain what you think will happen to the egg)

because ______________________________________

____.

(explain why you think this will happen)

Directions:

1. Gently place the egg in the container.
2. If you are using food coloring, pour some vinegar (around two ounces or 60 milliliters) into a glass and mix in food coloring until it is evenly mixed and the color you like. Pour the mixture over the egg. Add more vinegar until the egg is completely submerged. *If you are not using food coloring, pour vinegar into the container until the egg is completely submerged.*
3. Observe the egg and record your observations in the section below. Then set the container in an area where it will be safe. Leave it alone for 48 hours.

Draw what you see	Describe what is happening

*******Wait 48 Hours*******

Gently rinse the egg in water. Then complete the rest of this page.

Describe what happened	Was your hypothesis correct? Explain.	What kind of change occurred? Support your answer with evidence.

Balloon Experiment

Needed Materials:

- One standard size latex balloon
- One clean and empty plastic bottle
- White vinegar
- Baking soda
- A funnel
- Measuring tools

Predict:

In this experiment you will attach a balloon filled with baking soda to a bottle filled with vinegar. Make a hypothesis to predict what will happen to the balloon when the baking soda and vinegar meet in the bottle. You can use the provided sentence frame to help organize your thoughts.

My Hypothesis

If I pour baking soda into the bottle, then the balloon will ____

____

(explain what you think will happen to the balloon)

because ______________________________________

____.

(explain why you think this will happen)

Directions:

1. Measure one tablespoon (15 grams) of baking soda. Using a funnel, pour the baking soda into the balloon.
2. Measure 4 ounces (120 milliliters) of vinegar and pour it into the empty bottle.
3. Carefully stretch the lip of the balloon over the neck of the bottle, making sure the baking soda does not spill into the bottle.
4. Once you have the balloon sealed on the bottle, pinch your fingers around the area where the balloon and bottle meet to help keep the balloon secured in place.
5. Tip the contents of the balloon into the bottle.
6. Observe and then complete the rest of this page.

Draw how the balloon and bottle look before the change.	Draw how the balloon and bottle look after the change.

Was your hypothesis correct? Explain. ______________________________________

____

____

What kind of change occurred? Support your answer with evidence. ______________________

____

____

____

Hands-On Activities

Quick and Easy Ideas to Experience the Changes

Physical and chemical changes are all around us. There are many easy ways to demonstrate these changes, especially in a kitchen. For each change you experience with your student, be sure to discuss the reasoning behind why a change is considered physical or chemical. Below, you will find some ideas to get you started.

- Pop popcorn
- Use a bath bomb (or make them yourselves!)
- Make slime or homemade playdough
- Make an art project
- Give your student an object and see how many physical and chemical changes they can put it through
- Cook a meal together
- Bake and decorate a cake
- Melt broken crayons together to form one big rainbow crayon
- Light a candle or turn on a wax warmer
- Make bread dough, watch it rise, bake it, slather it with butter, and eat the bread together
- Make smoothies in a blender
- Make homemade popsicles
- Place dried beans on a damp paper towel and place it all in a sealed plastic bag to watch them sprout and grow over time
- Go on a nature walk and make note of all the changes you spot
- Make sweet tea, hot chocolate, or lemonade
- Use batteries to power a toy or flashlight
- Make a food that involves fermentation, such as kimchi or tempeh
- Play with glow sticks

Physical and Chemical Changes Poster Project

Guidelines & Grading Sheet

Directions:

Create a poster to demonstrate your understanding of physical and chemical changes. You can make your poster on a posterboard or use the blank poster sheet on page 42 of this workbook. Use the information below to successfully create your poster.

Poster Guidelines:

- The poster needs to be about physical and chemical changes.
- The poster needs to contain the definitions of physical and chemical changes.
- The poster needs to explain how to determine if a physical or chemical change occurred.
- The poster needs to list two examples of each type of change.
- The poster needs one fact about each type of change.
- The poster needs one picture for physical changes and one picture for chemical changes.
- All information needs to be accurate and show a good understanding of the topic.
- The poster needs to include a title related to the topic.
- The poster needs to have a decorative border around the outside.
- The poster needs to be neat, organized, creative, and colored.
- The poster needs to be entirely in the student's own words.

Poster Grading Sheet:

The poster contains the definitions of physical changes and chemical changes.	____/5
The poster explains how to determine if a physical or chemical change occurred.	____/10
The poster lists two examples of each type of change. (Four examples total)	____/5
The poster has at least one fact about each type of change. (Two facts total)	____/5
The poster includes one picture for each type of change. (Two pictures total)	____/5
Overall, the poster shows an accurate understanding of the topic.	____/5
The poster has a title related to the topic. The poster has a border.	____/5
The poster is neat, organized, creative, and colored.	____/5
The poster is entirely in the student's own words.	____/5

Total Score: /50

Physical Changes

Definition:

Examples:

- ○
- ○

Fact:

Chemical Changes

Definition:

How can you tell if a change is physical or chemical?

Examples:

- ○
- ○

Fact:

Picture:

Picture:

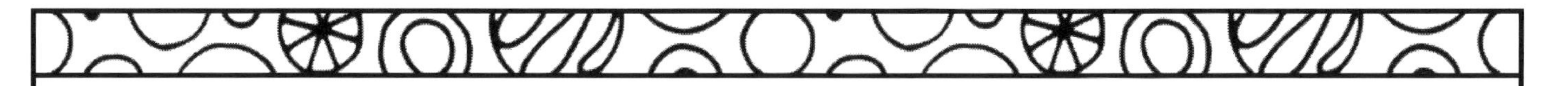

Physical and Chemical Changes Assessment

Directions: Read the statements below and determine if they are true or false.

1. True or False: A physical change makes changes at the atomic level.
2. True or False: Melting and condensation are both examples of physical changes.
3. True or False: Chemical changes always result in a color change.
4. True or False: It is usually easier to reverse a chemical change than a physical change.
5. True or False: Heating a metal bar until it turns red is an example of a physical change.
6. True or False: Smoke, temperature change, and light are signs of chemical changes.

Directions: Answer the questions below using complete sentences.

7. Is trimming a pet's fur a physical or a chemical change? How can you tell?

____
____
____
____

8. Is a bicycle rusting in the rain a physical or a chemical change? How can you tell?

____
____
____
____

9. Is painting a wooden fence white a physical or a chemical change? How can you tell?

____
____
____
____

10. After mixing two liquids together you notice a new sandy substance on the bottom of the container. Did a physical or a chemical change occur? How can you tell?

____
____
____
____

Answer Key

Physical and Chemical Changes Overview (Page 10)

1. A physical change is a change that does not change the identity of the substance.
2. *Answers will vary. Example answers:* Cutting an apple and sanding a piece of wood
3. A chemical change is a change that changes the identity of the substance and creates something new that was not there before.
4. *Answers will vary. Example answers:* Frying an egg and bread baking
5. Changes in states of matter are physical changes because they are reversible and nothing new is created.
6. Some signs a chemical change occurred are the creation of smoke, gas, heat, or light. Color changes, a new smell, and the formation of a precipitate are also signs of a chemical change.
7. *Possible questions include the following:* Do I have something new that was not there before? Is the change reversible?

Chart answers will vary. Possible physical changes of the paper include cutting, folding, crumbling, tearing, and coloring. Possible chemical changes of the paper include burning and eating.

Physical and Chemical Changes Makeover (Page 11)

Answers will vary. Possible physical changes include a haircut of some sort, curling hair, blow-drying hair, painting nails, filing nails, adding extensions, putting on lipstick, etc. Possible chemical changes include bleaching hair, getting a perm, relaxing hair, etc.

Physical and Chemical Changes Examples (Page 12)

1. *Circle* "plate breaks in half"
2. *Underline* "leaves change color"
3. *Underline* "burns toast"
4. *Circle* "condensation"
5. *Circle* "mix flavored drink powder with water"
6. This is a chemical change. The donut goes through the digestive process. It...ahem...doesn't smell or look like a donut after that chemical change. Those rainbow sprinkles aren't exactly the bright sugary treats they used to be.
7. This is a chemical change. The color change in the strawberries indicates a chemical change. The composition of the strawberry is changing at the atomic level. The ripening of the strawberry cannot be reversed.
8. This is a physical change. The aluminum can is still aluminum. It just looks and feels different. It is possible to reverse this change by melting and reshaping the can.
9. This is a chemical change. The milk has a new smell. Gas is released (that is why the carton puffs out a little). Milk chunkies (yum) show a precipitate has formed. The change is not reversible.
10. This is a physical change. The popsicle went from a solid to a liquid, which is a change in state of matter. Changes in states of matter are physical changes that are reversible. It is possible to collect the popsicle liquid and freeze it to have a popsicle again.

Answer Key

Physical and Chemical Changes Identification (Page 13)

1. Fireworks exploding is a chemical change because the fireworks produce light and smoke. There is also a color change and a new smell. The change is not reversible.
2. Making paper snowflakes is a physical change because nothing new is created. The paper looks and feels different, but it is still paper. No signs of a chemical change occurred.
3. Digesting food is a chemical change because something new is created. Gas is produced. The food smells different and becomes a new color.

Physical and Chemical Changes in Your Daily Life (Page 14)

****Student answers will vary. Answers below are examples.****

In a garden:

Physical Change: Mixing soil. This is a physical change because you are just combining materials together. Nothing new is created that was not already there.

Chemical Change: Plants going through photosynthesis. This is a chemical change because it involves light being changed into energy and the creation of oxygen.

At a dentist's office:

Physical Change: Drilling into a tooth. This is a physical change because the tooth is still a tooth. It has a different shape and texture.

Chemical Change: Whitening teeth. This is a chemical change because a color change occurred.

In a restaurant:

Physical Change: Chopping up vegetables. This is a physical change because the vegetables stay vegetables. They just change size.

Chemical Change: Cooking a hamburger. This is a chemical change because the hamburger changes colors and makes a new smell.

On a beach:

Physical Change: Building a sandcastle. This is a physical change because the sand is still sand. It just looks different.

Chemical Change: Having a bonfire. This is a chemical change because the wood turns into something else. Heat, light, & smoke are made.

At a movie theater:

Physical Change: Ice melting in your cup. This is a physical change because melting is a change in state of matter and can be reversed.

Chemical Change: Popping popcorn. This is a chemical change because it is irreversible. A sign of the change is the new smell of the popcorn.

Answer Key

Physical and Chemical Changes Write or Draw (Page 15)

****Exact answers will vary and may include pictures and/or drawings to represent answers.*

1. A physical change is a change that does not change the identity of the substance.
2. A chemical change is a change that causes the identity of the substance to change.
3. Yes, it is possible for an object to go through a physical and a chemical change at the same time. An example is a candle. While the candle is lit, the wax melts (physical change) and the wick burns (chemical change).
4. Some ways a physical change might be reversed are changing a temperature (a melted object can be cooled to return to a solid form), filtering or sorting by hand to remove particles, and evaporating (heating salt water turns the water into vapor, leaving the salt behind).
5. The particles should be grouped the same but rearranged in the space.

6. The particles should be grouped differently.

7. The production of light, heat, gas, smoke, a color change, a new smell, or a precipitate
8. Physical changes do not make something new, while chemical changes result in something that was not there before. Physical changes can often be easily reversed but chemical changes cannot.

Physical and Chemical Changes Odd One Out (Page 16)

****Answers will vary. It is possible for a question to have more than one correct answer. The following are example correct answers.*

1. Something new is created. This is the odd one out because it is a chemical change. The other three are facts about physical changes.
2. Breaking a glass. This is the odd one out because is an example of a physical change. The others are examples of chemical changes.
3. A change in shape. This is the odd one out because it can be a sign of a physical change. The other three are signs a chemical change occurred.
4. The identity of the item is the same as before. This is the odd one out because the item in a physical change has the same identity it did before the change. The others are about what can happen in a chemical change.
5. Making water by joining oxygen and hydrogen atoms. This is the odd one out because atoms joining together to make something new is a chemical change. The other three involve changing state or dissolving, which are physical.
6. A paper turns from white to red with paint. This is the odd one out because painting something does not change the identity of a substance. It is a physical change. The other changes in color are changes on a chemical level.

Answer Key

Card Sorting Activity (Page 25)

- *Physical Changes:* A change that does not change the identity of a substance; it is still the same but might look or feel different, Melting ice, Breaking glass, Cracking an egg, Boiling water, Slicing bread, Making lemonade, Mowing the lawn, and Breaking a pencil
  - *Answers may vary. I can tell these are physical changes because...*Nothing new was created in these changes. Some of the changes were reversible and some were changes in states of matter.
- *Chemical Changes:* A change that changes the identity of a substance; something new is created that was not there before, Lighting a match, Baking bread, Making toast, Shooting off fireworks, Digesting food, Roasting marshmallows, Rusting nail, and Frying eggs
  - *Answers may vary. I can tell these are chemical changes because...*Signs of chemical changes like smoke, color change, and new smells happened with these changes. The changes were irreversible. Something new was created in each of these changes.

Answers for questions 1, 2, and 3 will vary.

Stations Activity (Pages 27-35)

1. A physical change is a change that does not change the identity of the substance.
2. A chemical change is a change that causes the identity of the substance to change.
3. False. Physical changes can often be reversed, not chemical changes. In chemical changes the item is no longer the same thing it was before the change, so it is very difficult or impossible to reverse the change.
4. True. Physical changes include changes in states of matter because the item is still the same as it was before the change. Water is still water whether it is melted or evaporated or frozen.
5. Answers may include: light, heat, smoke, the production of gas, a color change, a new smell, or the formation of a precipitate
6. Stirring sugar into tea, Mixing brownie batter, and Spreading butter on bread
7. A lake freezing and Chopping down a tree
8. Possible drawing ideas include photosynthesis, decaying, and burning.
9. Possible drawing ideas include cutting, trimming, replanting, and transpiration.
10. Possible answers include eating it, roasting it, setting it on fire, or letting it rot.
11. Possible answers include peeling it, chopping it, dipping it in ranch dressing, or mixing it into a salad.
12. Rising bread dough in a bowl
13. *Answers will vary.* Possible answers include turning on a light, eating pizza, and mixing an acid with a base.
14. *Answers will vary.* Possible answers include freezing water, popping a balloon, and painting a fence.
15. Answers include popping popcorn, the bag of popcorn smoking and on fire, Lady eating the popcorn, and Lady's toots
16. Answers include cutting paper, coloring the heart, gluing pompoms, folding the heart, and stuffing the heart into an envelope
17. Possible answers include the formation of gas bubbles in the cake making it rise, the darkening color of the cake, and the delicious smell coming from the oven
18. Possible answers include the pumpkin changing colors, the gas that is emitted from the rotting, and the bad smell

Answer Key

Poster Project (Pages 41-42)

Example to the right.

Assessment (Page 43)

1. False
2. True
3. False
4. False
5. True
6. True
7. Cutting a pet's fur is a physical change because the fur only looks and feels different. Nothing new is created.
8. A bike rusting in the rain is a chemical change for many reasons. Something new is created that was not there before (rust). There is a color change and a slight smell.
9. Painting a fence is a physical change. Although there is a color change, nothing new is created and the color change is only a result of placing a color on top of wood. It is still just a wooden fence and paint; it only looks and feels different. The paint can be removed from the fence, so the change can be easily reversed.
10. A chemical change occurred. When the two liquids mixed together, they formed a precipitate. The precipitate is a new substance that was not there before.

Differences Between Physical & Chemical Changes

Physical Changes

Definition:

A physical change is a change that does not change the identity of the substance. The item is still itself; it only looks or feels different.

Examples:

- *Boiling water*
- *Tearing paper*

Fact:

Changes in states of matter, like freezing and melting, are always physical changes. If you make ice cubes in the freezer, you still have water. It is just in a solid form.

Picture:

Breaking a pencil is a physical change. The pencil is still a pencil; it just looks different.

Chemical Changes

Definition:

A chemical change is a change that makes the actual material in a substance change. The material is not the same thing anymore; it is something new.

Examples:

- *Digesting food*
- *Shooting fireworks*

Fact:

Even in a chemical change, no new atoms are created and no existing atoms are destroyed. The atoms are just rearranged and regrouped.

Picture:

Frying an egg is a chemical change. Signs of the chemical change include a color change and a new smell.

How can I tell if a change is physical or chemical?

A physical change does not create anything new. Physical changes are often reversible. A chemical change will make something new that was not there before. There are several signs that a chemical change occurred. Those signs include light, heat, smoke, gas, a color change, and a new smell.

About the Author:

Elly Thorsen attended the University of South Dakota where she earned her BS in Elementary Education and MA in Technology for Education and Training. She has taught in South Dakota, Minnesota, Oklahoma, and South Korea. She has experience teaching many grade levels and subject areas in a variety of school settings but loves seventh grade science best.

Find more educational resources for your student!

https://www.amazon.com/author/ellythorsen

https://www.teacherspayteachers.com/store/elly-thorsen

Made in the USA
Las Vegas, NV
12 December 2024